The Botanical Universe

The Botanical Universe

ANALOGIES OF GARDENS AND GALAXIES

. . .

frank louis quattromani

ISBN: 1537183613
ISBN 13: 9781537183619
Library of Congress Control Number: 2016913989
CreateSpace Independent Publishing Platform
North Charleston, South Carolina

The Botanical Universe
(Analogies of Gardens and galaxies, teaching-by-analogy)

• • •

THE UNIVERSE IS A VAST and complex place, with as many as two trillion galaxies. It can be difficult, even intimidating, to try to grasp complicated astronomical and cosmological concepts. However, with analogies comparing the unknown universe with known ordinary trees, plants and birds in the garden, we can unlock some of the complex mysteries and secrets of our universe. Utilizing the known world of botany we can explain the unknown world of astronomy and cosmology. "Carefully crafted, elaborate science analogies can help us build conceptual bridges between what we already know and what we are setting out to learn (Glynn)." Understanding how a giant sequoia tree grows from a single tiny seed can help us understand how our universe started with the "Big Bang" from a single minute nidus of matter. Understanding how birds in flight can pull each other along can help us conceptualize Einstein's theory of general relativity with warping of space/time.

"An analogy is a comparison of the similarities of two concepts. The familiar concept is called the analog and the unfamiliar one the target (Glynn)." In an article by Shawn Glynn, the familiar or analog is the tree, flower and bird and the unfamiliar is the universe. Analogies can provide us with a unique perspective on the connections between the largest and the smallest systems in the universe." Analogies often begin with …think of it as…(Glynn)."

Shawn M Glynn in an excellent article, *making science concepts meaningful to students: teaching by analogies*, has demonstrated that "science-education research studies and science teachers' classroom experiences utilizing analogies can help make science concepts meaningful to students. The article describes how Johannes Kepler (German astronomer-circa 1600), used analogies to help explain his work by comparing the celestial machine with clockwork. Every learning process includes a search for similarities between what is already known and the new, the familiar and the unfamiliar…."

Zen master Thich Nhat Hanh wrote, *"a grain of rice contains the universe… when we look at a grain of rice, we see that this grain contains the whole world, the rain, the clouds, the earth, time/ space, farmer, everything."*

Preface

. . .

GALILEO GALILEI (ASTRONOMER, CIRCA EARLY 1600) and **DR. GEORGE WASHINGTON CARVER** (botanist, circa early 1900) meet somewhere in the great beyond and marvel at the universe below. As they converse they begin to realize how much the world of astronomy, cosmology and botany have in common, cosmology being a branch of astronomy that deals with the origin and evolution of the universe. They talk about how the laws of gravity, energy, relativity, singularity, harmony, matter, and function apply equally to astronomy, cosmology, botany and birds in flight.

GALILEO GALILEI talks about the "big bang" theory and how our current universe began about 13.8 billion years ago from the expansion of a minute nidus of matter. He recalls when Carl Sagan proposed the concept of the "Cosmic Calendar" as a method of visualizing the chronology of the universe, scaling its current age of 13.8 billion years to one calendar year. By this scale there are 37.8 million years per day. If the Big bang occurred on January 1 at midnight, the first galaxy would have formed on January 22, and the planet earth would have coalesced from "star stuff" by September 1. Photosynthesis on earth would have begun on September 30 and the first flowers seen on December 28. Modern history occurred during the last second of the year, December 31, at 23:59:59 PM.

DR. GEORGE WASHINGTON CARVER thinks of that moment in time perhaps several hundred million years ago on planet earth when the first molecule of plant life gave origin to all existing plant life today. A sort of botanical "Big Bang." Dr. Carver then draws attention to the single tiny seed from which a giant sequoia tree will grow.

Author

• • •

Frank Quattromani, MD. Clinical Professor of Radiology
Pediatric Radiologist, University Medical Center and Texas Tech
University, Lubbock, Texas
United States Army Retired
Radiology Consultant to the Surgeon General

CONTRIBUTING AUTHOR
Patricia Hull McEachern Quattromani (amateur botanist)

SUPPORT
Eman Attaya, MD (Cover illustrator and editing)
Stephen Frazier, MD (Camera support)
David Cervantez (Technical support)
Max Quattromani (Photoshop and technical support)

Dedication

• • •

To the FAM: My dad, Dr. Frank, and mom, Josephine, Libby, Scott, Jeff, Max and Jim.

and my wonderful daughters in law and beautiful grandchildren.

To Pat for her support and encouragement.

Acknowledgments

• • •

It would be impossible to write a book about astronomy and botany without recognizing those organizations both governmental and private that led the way in research and development.

I am indebted to Kahn Academy; Smithsonian.com; Carl Sagan; St. Louis Botanical garden; Ask the Astronomer.com; NASA.com; Hubble.com; Astronomy Café; Smartscience pro; Scienceblogs NASA/ ESA; Pinterest; Alan Duffy abc.net.au; Pearson Education. Inc; Valberta.Ca; rampages.us; NASA-Jeff Hester; The Physics of the Universe; Sciencemag; Vanessa Janek; Michio Kaku-Cosmicstream; Project Noah; Delange; Frank Lorey; Mialphaniomega.wordpress; Dreamstime; Pics-about-space; Isaac Antwi; Astronoteen; Trumpetflowers.com; Keoneulaes.org; Sciencephoto library; Ron Kurtus; Astronomy links; Astronomy Picture of the Day; The Universe:Quantum microscopic universe; The dark side of the Universe; deskeng.com; macsmotorcitygarage.com; solidworks.com; wordpress.com; wikiwand (Banksia flower); universe-review;caabc.net.au; wiseGEEK. writescience.wordpress.com; science20.com; crystalinks.com; fotothing.com; sweetpics.site; imagejuicy.com; onlineplantguide.com; thisfabtrek.com.Topix.com (yellow trumpet flower)Marklemer.wordpress (constellation Orion), pixabay.com (sequoia tree)Krysia Zajonc (localfoodlab.com (Zen image), glogster.com (George Washington Carver)

Justus Sustemans(Wikipedia), Big Bang (fanboy.com) Carolinanature. com (Flower) dewharvest.blogspot(hand with seeds. writescience.info (diagram black hole), fun.putidea.info (wine-cup flower)Pinterest.com(wine-cup flower). Waykiwayki.com (special relativity gravity), radiolopolis.com (drooping lily), quora.com (birds in flight), solar.physics.montana.edu(rose nebulus) space.com (dark matter) Celosia flower (pinterest). mediawiki. sites.thefulwiki.org. Comet Hale Boop (Detlev Van Ravenswaay) Shawn Glynn

Every effort has been made to acknowledge owners of images, however we apologize if there are any unintentional omissions.

Contents

CHAPTER I

The Botanist and the Astronomer
(The ultimate scientists)

• • •

DR. GEORGE WASHINGTON CARVER (BOTANIST, inventor, man of faith, educator and humanitarian) was born into slavery in 1864 in Diamond Grove, Missouri and died on January 5, 1943. Carver's reputation as a botanist is based on his research into and promotion of peanuts and sweet potatoes for added nutrition for poor farm families. He was a leader in promoting environmentalism. In 1896, Booker T. Washington invited Dr. Carver to head the college agricultural department at Tuskegee Institute where he published 44 practical bulletins for farmers. Carver taught there for 45 years. He taught methods of crop rotation and was first to understand the necessity of crop rotation to replenish the soil with nitrogen. Dr. Carver developed a peanut oil massage to treat polio and contributed to our knowledge of alfalfa, wild plum, the tomato plant and corn as well as peanuts. Dr. George Washington Carver was commemorated with a three cent stamp in 1948 and featured on a silver dollar coin with Booker T Washington minted in Philadelphia in 1952. The two men are buried together at Tuskegee University in Alabama. On his grave is written, *"he could have added fortune to fame, but caring for neither, he found happiness and honor in being helpful to the world."*

GALILEO GALILEI (astronomer, physicist and mathematician) was born on February 15, 1564 and died January 8, 1642. He was an Italian astronomer, physicist, engineer and mathematician and is called the father of observational astronomy. Galileo defended heliocentrism in the Catholic court believing the earth rotated around the sun. Galileo was considered a heretic by Pope Urban VIII and convicted by 6 of ten inquisitors in a formal ceremony at the church of Santa Maria in Siena, Italy. He was placed under house arrest by Pope Urban VIII. He remained under house arrest despite many medical problems until his death in 1642. The church finally accepted that Galileo might be right concerning heliocentrism in 1983.

GALILEO GALILEI

ASTRONOMER

CHAPTER 2

Big Bang and the sequoia tree (Where it all began)

• • •

IF GALILEO AND DR. CARVER were to converse, this is how it might sound.

GALILEO GALILEI, "What a beautiful cosmos we have and to think it all began with a "Big Bang" from a single point in space. More than thirteen billion years ago a minute nidus of matter expanded space giving birth to our universe. The expanded space of the Big Bang was filled with hot gases that ultimately coalesced to form galaxies filled with millions of stars. That expansion continues today and at an accelerated pace."

GEORGE WASHINGTON CARVER, "Daily on earth, small plant seeds begin absorbing energy and nutrition from the surrounding ground soil and grow into giant trees like the Sequoia tree or tree groves such as the aspen grove. The giant sequoia tree (General Sherman) in California grew from a single small seed. The aspen trees which are all connected to a single root system have their origins in the first single plant molecule which began with the planet earth's botanical "Big Bang" over one billion years ago. Just as everything that the sequoia tree will ever be is stored in that tiny sequoia seed so the entire universe was once concentrated is a single point, the singularity. If the analogy comparing the growth of the sequoia tree and the origin of the universe is in fact close to reality, then the "Big"Bang" universe may not have begun with such a chaotic explosion but rather a more planned, orderly, purposeful and controlled expansion.

ARTIST CONCEPT OF THE BIG BANG

BANKSIA SERATA FLOWER WITH CENTRAL OVARY,
LONG FILAMENTS AND ANTHER ANALOGOUS TO THE
COSMIC "BIG BANG" WITH GALAXIES FLYING OUTWARD.

CEPHALANTHUS OCCIDENTALIS FLOWER

WITH ITS RADIATING FILAMENTS AND ANTHER

RESEMBLES THE COSMIC "BIG BANG."

Sequoia seeds

GENERAL SHERMAN

CHAPTER 3

Black Hole Complex and the trumpet vine flower (Dante's inferno-abandon all hope, ye who enter)

. . .

GALILEO GALILEI, THERE ARE FOUR components to the black hole complex:

1) Hidden in the center of the black hole complex is the gravitational singularity, a region of space where the density of matter and the curvature of space/time becomes infinite.
2) The gravitational singularity is surrounded by the black hole, where gravity is so strong that nothing can escape, not even light.
3) At the outer limit of the gravitational effect of the black hole is the event horizon.
4) Just beyond the event horizon is an accumulation of matter such as gas, dust and other stellar "star stuff" (Carl Sagan) in the form of a flattened band of orbiting matter (accretion disk) traveling at almost the speed of light."

DR. GEORGE WASHINGTON CARVER, "Deep inside that *Trumpet Vine* (*Campsis radicans*) flower is a nidus (singularity) of sugar-rich nectar so sweet with such a strong invisible scent that it is irresistible to the hummingbird. Like the pull of gravity by the singularity of the black hole, the scent of the sweet nectar of the *Trumpit Vine* flower attracts the hummingbird."

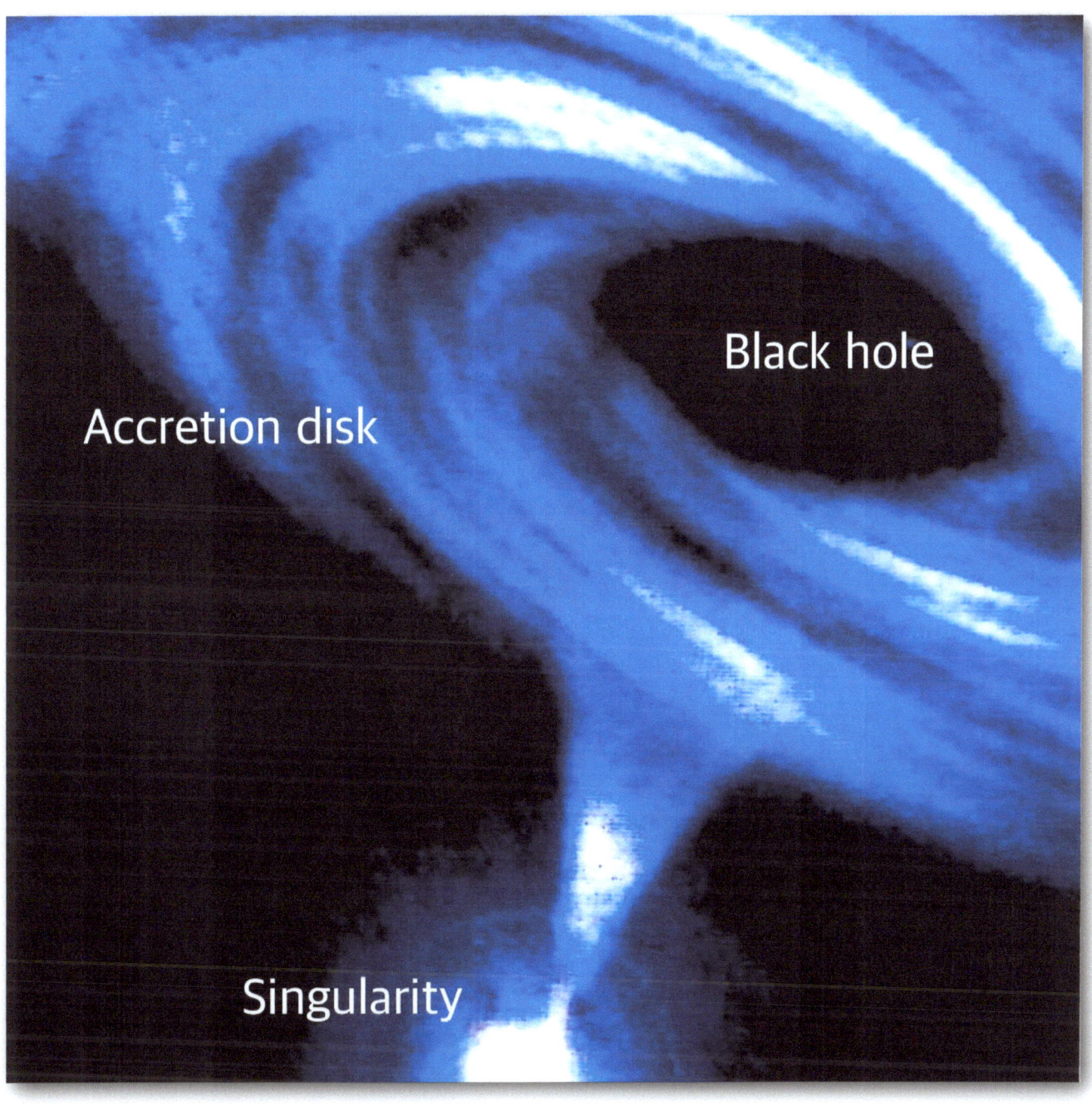

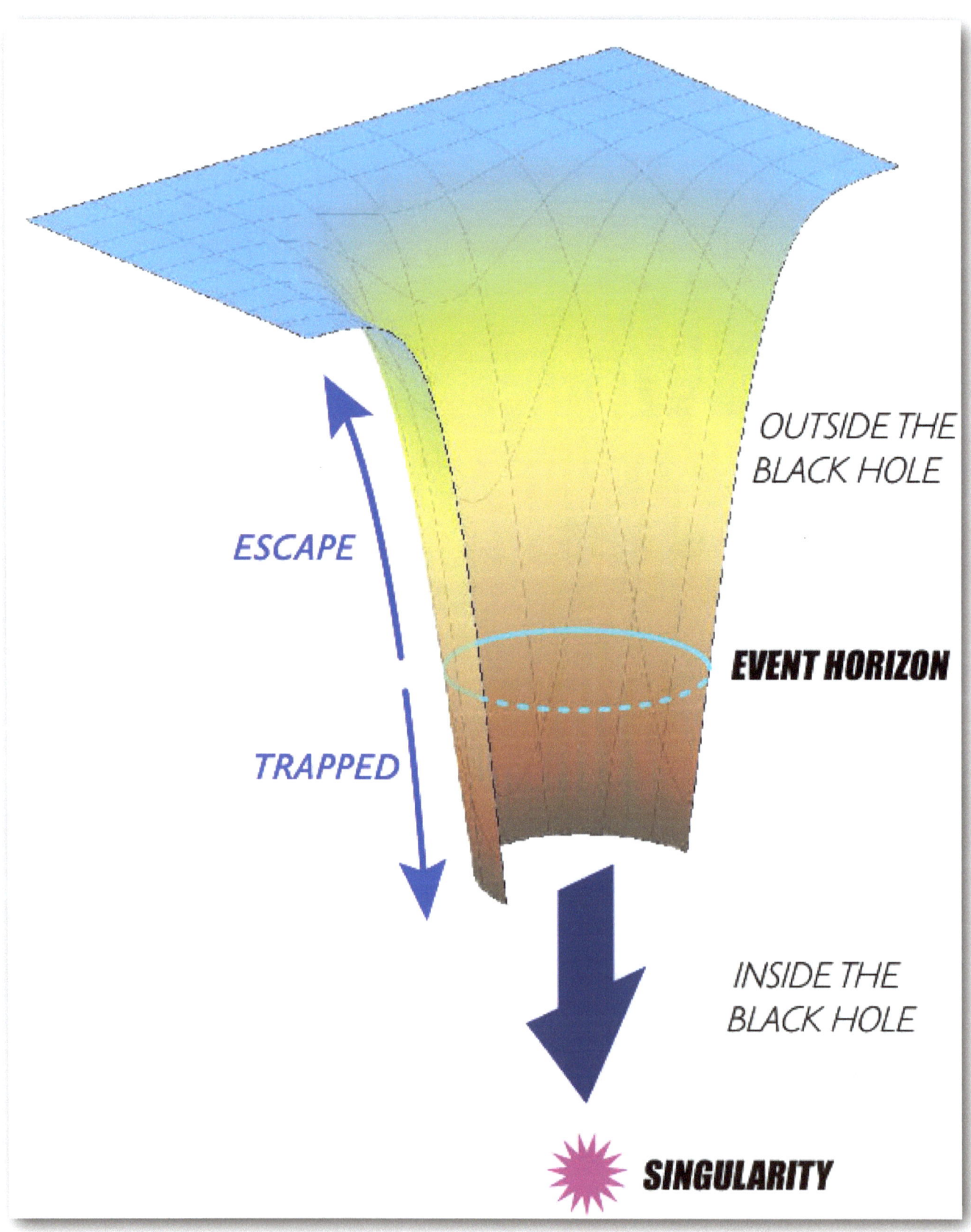

WINE-CUP FLOWER WITH THE BLUE (ACCRETION DISK), WHITE (EVENT HORIZON) RED-BLACK (BLACK HOLE) AND GREEN SINGULARITY

Wine cup flower with accretion disk (blue), event horizon (white), black hole (red–black) and singularity (green).

CHAPTER 4

Dark Energy expanding the cosmos compares to water and products of photosynthesis growing and expanding the tree.

• • •

GALILEO GALILEI, "DARK ENERGY PERMEATES all space uniformly and accounts for 68.1% of the total /mass energy in the universe. Dark energy is responsible for the accelerated expansion of our universe. All galaxies are moving away from us, with the more distant galaxies receding faster. Dark energy exerts a negative, repulsive effect, behaving like the opposite of gravity, and providing a way of balancing the gravitational contraction caused by Dark Matter. Research thus far suggests that Dark Energy and Dark Matter are two separate and distinct forces acting totally apart from each other with Dark Energy permeating all space uniformly creating expansion of the universe while Dark Matter functions as a completely separate and independent non-uniform cosmic force for gravitational contraction. The vacuum expansion effect of dark energy on the cosmos would compare to the osmotic effect of water and nutrients moving up the tree for tree growth and expansion.

DR. GEORGE WASHINGTON CARVER, "Trees, like the universe, require energy to grow and expand. Energies necessary for tree growth includes water and products of photosynthesis. Energy necessary for growth of the universe is called Dark Energy.

Just as Dark Energy accounts for two-thirds of the mass /energy of the universe so water and products of photosynthesis constitute two-thirds of the mass of the tree.

Perhaps the true nature of Dark Energy will be clarified at the conclusion of the HETDEX project The HETDEX/VIRUS project will begin in late 2016 at the McDonald Observatory on Fort Davis Mountain in south Texas.

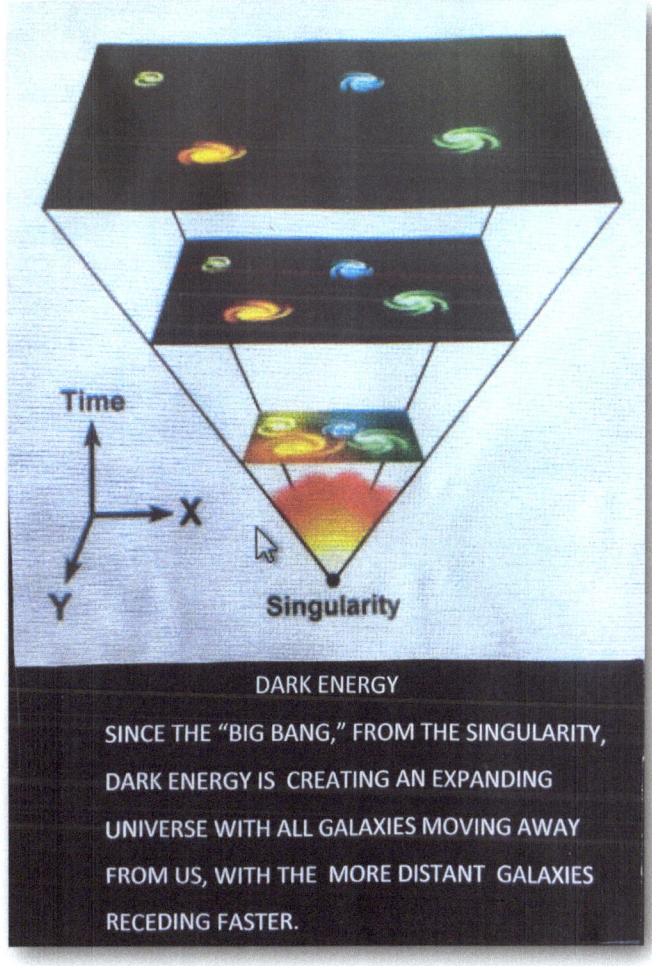

Dark Energy (energy to the tree)
Energy to the tree derives from water up from the roots and
photosynthesis down from the leaves.

CHAPTER 5

Dark Matter and the tree trunk (Dark Matter and the tree trunk hold it all together)

• • •

GALILEO GALILEI, "Dark matter is a mysterious force that accounts for 27% of all matter in the cosmos. The other 68.1% and 4.9 % are dark energy and ordinary matter respectively. Dark Matter is denser where galaxies cluster, in fact the existence and properties of Dark Matter is inferred from its gravitational effects on visible matter such as large galaxy clusters. Dark Matter, like gravity, holds the cosmos together. If it were not for the gravitational effect of dark matter, stars in the outer rings of galaxies at the outer margins of the universe would fly off out into deep space.

As an analogy, we will compare the tree trunk with Dark Matter**.** The tree trunk provides stability to the tree just as cosmic Dark Matter provide stability to the universe. The trunk of the tree holds the branches and leaves in place while cosmic Dark Matter holds the galaxies and gasses together. "

DR. GEORGE WASHINGTON CARVER, "The tree trunk has the duel function of both providing for stability to the tree while also transporting water and products of photosynthesis for expansion and growth of the tree. In an article in Biology4Kids.com there is a description of how plants evolved to be larger by developing a kind of circulatory system. The xylem of the tree trunk is a system of tubes and transport cells that carry water and nutrient from the soil to the branches hundreds of feet high above the ground. The tree trunk holds the tree steady while the xylem system in the tree trunk transports the energy of water and nutrients for expansion. At the end of the growing season the tree xylem becomes quiescent to ultimately take its place as the next seasonal tree ring that accumulates one per year."

The tree trunk has a two-fold function, primarily for the stability of the tree and secondarily to provide the framework for the transport of water and photosynthesis for growth. Perhaps in the cosmos, the dark matter which is responsible for stability of the universe is also the framework along which dark energy travels.

GALAXIES

DARK MATTER

Artist conception of clusters of bright galaxies held together by black branches of dark matter.

CHAPTER 6

Galactic rings and the tree rings
(Where time is recorded)

• • •

GALILEO GALILEI, "Galactic rings are a spiral collection of stars, the oldest stars of which are found in the center of the galaxy. Spiral galaxies are categorized according to the tightness of the rings. The inside-out theory of galactic growth is well documented. Initially, there is rapid star formation which begins in the galactic center. This is followed later by star formation in the outer rings of the galaxy. Star formation starts at the core of the galaxy and spreads outward."

DR. GEORGE WASHINGTON CARVER, "Tree rings like the rings of the galaxy also develop from inside-out. The oldest rings are in the center with subsequent years adding rings just under the outer bark.

The cold years in Europe between 1645-1715 was a period of reduced solar activity with low temperatures which caused a reduction in the diameter of the tree and as a result tight spiral tree rings. The colder the winter the tighter and closer the tree rings. The cold environmental changes to the tree in Europe in the 17th and 18th centuries produced tight rings in the Spruce tree which when used to make violins, produced a superior quality of sound that was the Stradivari's golden period.

CUT TREE SHOWING RINGS WHICH RESEMBLE THE GALAXY RINGS. THE RINGS OF THE TREE AND GALAXY GROW FROM INSIDE-TO OUT.

CHAPTER 7

Nebulae and the cactus plant
(Where stars are born, live and die)

• • •

GALILEO GALILEI, "A COSMIC NEBULA is an amorphous collection of interstellar gas and dust that can condense and give birth to stars. The dust consists of microscopic particles of silicates, carbon, ice, hydrogen, oxygen, helium and nitrogen. Nebulae of note in the cosmos include the Orion nebula in the Orion constellation, the rosebud nebula, the Pillars of Creation in the Eagle Nebula and the Rose nebula, all of which are star nurseries."
"Nebula stars form from the gravitational collapse of gas in the interstellar medium, gas from a supernova explosion. The Orion nebula is the brightest nebula in the sky. The first true nebula was discovered in 964 by the Persian astronomer Abd al-Rahman al-Suf. The supernova that created the Crab nebula was first observed by Arabic and Chinese astronomers in 1054."

DR. GEORGE WASHINGTON CARVER, "The amorphous matter of the *Opuntia Monacantha cactus* plant gives birth to a reddish-purple, fleshy, conical starlike fruit.

The Rosebud nebula is analogous to the *Celosia* flower and the Rose nebula resembles the rose."

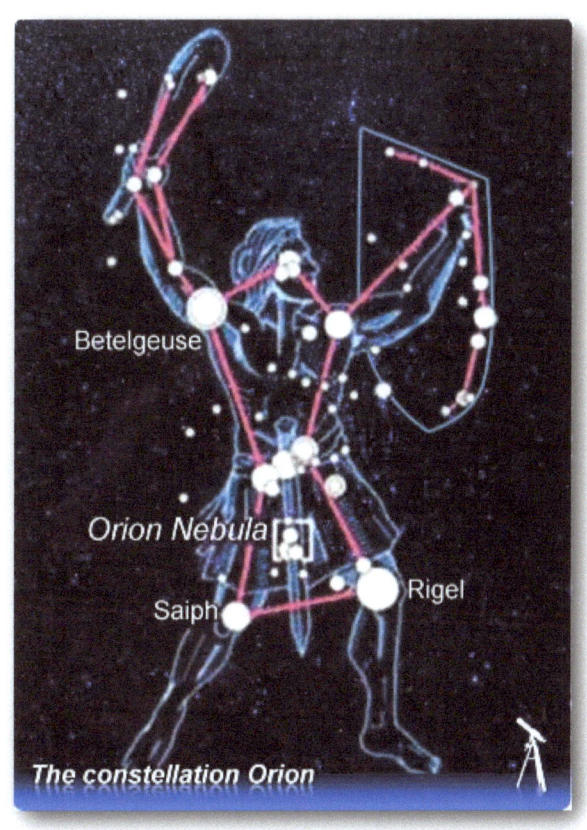

ROSEBUD NEBULA
RESEMBLES THE CELOSIA
FLOWER IN BLOOM.

THE CELOSIA FLOWER IN BLOOM WITH
THE SAME PATTERN AS THE ROSEBUD NEBULA.

Stars forming in the Pillars of Creation in the Eagle Nebula imitates the same configuration as the flowers born in the Opuntia Monacantha cactus plant.

Indian fig opuntia cactus plant with red fruit flowers growing from its body. This imitates the way stars are born in the Pillars of Creation in the Eagle Nebula

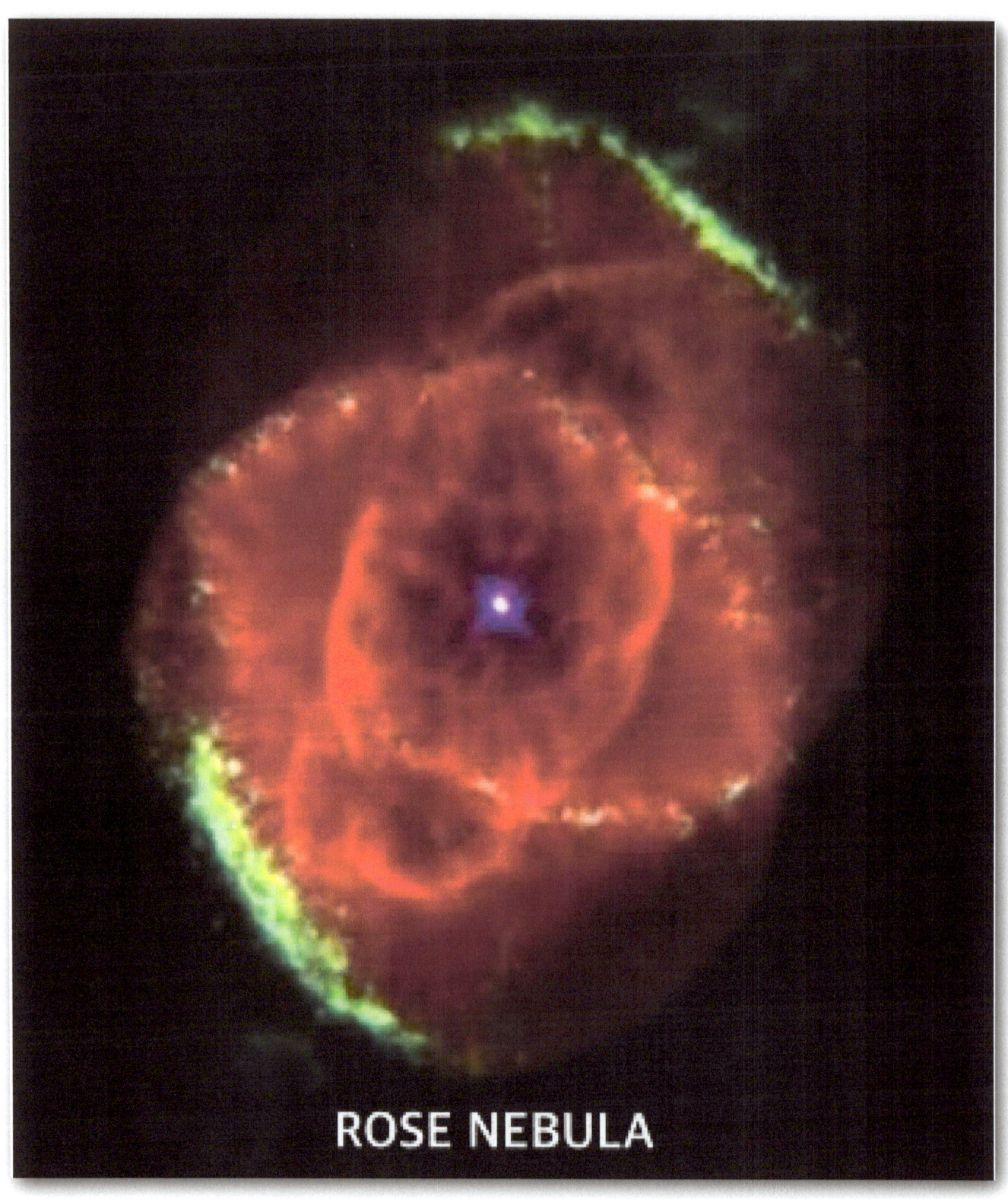

ROSE NEBULA

CHAPTER 8

Planet and fruit cores (Both planets and fruits have cores)

• • •

GALILEO GALILEI, "Planets have a crust, mantle and core.

The planet Mercury has a core that is largely iron sulfate and nickel and is very hot. The core of Mercury occupies almost 85% of the planets volume."

The planetary core consists of the innermost layers of a planet, which may be composed of solid and liquid layers. In our solar system, core size can range from a radius of about 20% (core of our moon) to 85% (core of Planet Mercury).

DR. GEORGE WASHINGTON CARVER, "The Avocado and peach fruits also have central cores or pits that similarly make up a significant portion of the fruit."

ARTIST'S DRAWING OF THE PLANET MERCURY
WITH ITS CRUST AND CORE CAN BE COMPARED
TO MANY FRUITS WITH A PEEL AND PIT
SUCH AS THE PEACH AND THE AVOCADO.

THE SLICED PEACH WITH ITS PEEL AND PIT STRONGLY RESEMBLES THE PLANET MERCURY WITH ITS CRUST AND CORE.

Avocado Core

The sliced avocado fruit with
its thick peel and large pit is very
similar to the planet mercury with
its thick crust and dense iron core.

CHAPTER 9

The Comet and the flying tree seed (Comets carry water while seeds carry genetic material to replicate the tree)

• • •

GALILEO GALILEI, "The origin of the earth's oceans remain a hotly debated topic but one of the leading hypothesis is that most of the water came from comets or asteroids. The comet is an orbiting solar system body which when heated by the sun can give off both gas and dust in the form of two tails, a gas tail and a dust tail. Both tails are always directed away from the sun. The gas (ion) tail pushed by solar winds points straight away from the sun. The dust tail follows the orbital path of the comet. Comets bombarding the young earth about 4.6 billion years ago brought vast quantities of water to the earth. Even molecular oxygen has been found in the comets."

DR. GEORGE WASHINGTON CARVER, "The Maple tree seed also has two tails permitting it to be aerodynamic with a helical motion when carried by the wind. Maple tree seeds (fruits) are called samara. Each seed is contained in a "nutlet" attached to flattened wings. Maple tree seeds contain the genetic material including the amino acids necessary to replicate the tree from which it fell."

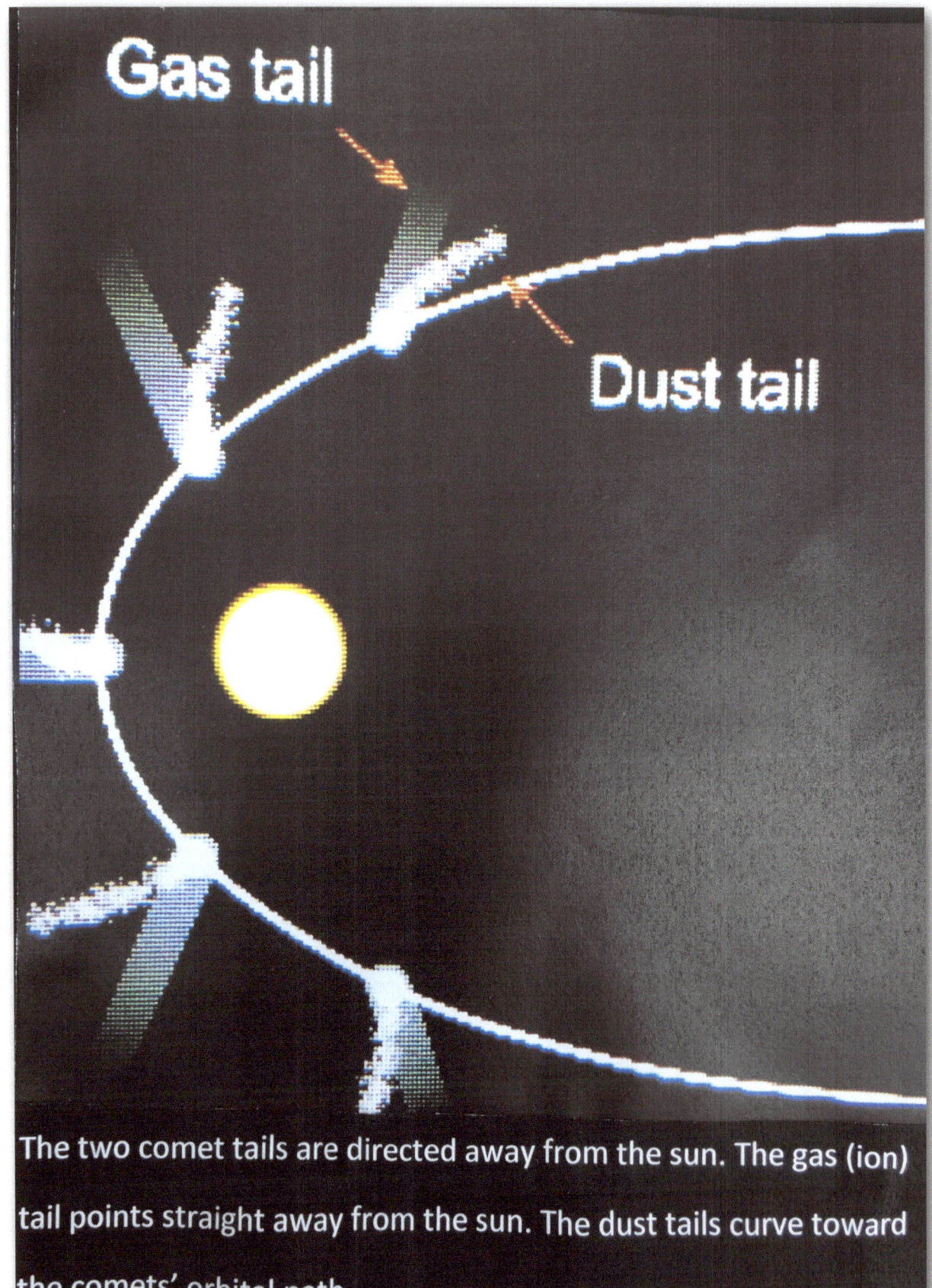

The two comet tails are directed away from the sun. The gas (ion) tail points straight away from the sun. The dust tails curve toward the comets' orbital path.

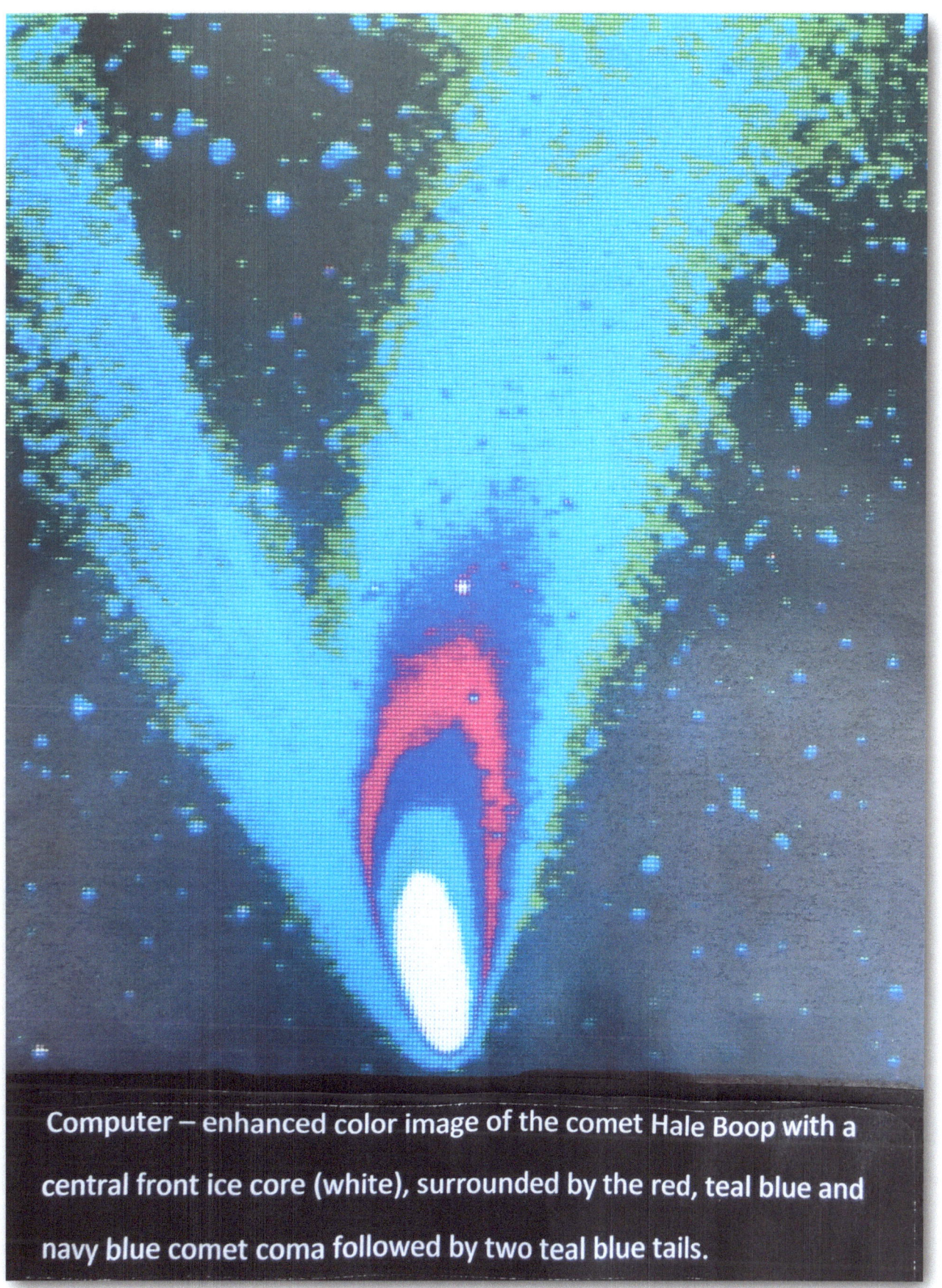

Computer – enhanced color image of the comet Hale Boop with a central front ice core (white), surrounded by the red, teal blue and navy blue comet coma followed by two teal blue tails.

MAPLE TREE WITH
"COMET-LIKE" SEEDS

Gravity and the drooping lily (How one mass in motion can affect the motion of adjacent masses)

• • •

GALILEO GALILEI, "GRAVITY IS WHAT makes pieces of cosmic matter clump together into planets, stars and galaxies. Gravity is what keeps the earth in orbit around the sun. Gravity makes the trees on the surface of the earth grow perpendicular to the earth's core and not the earth's crust. Gravity for the planet earth is centered at the earth's core and decreases with distance from the core. Yes, trees sense gravity! Stars in the outer rings of galaxies do not fly off away from the parent galaxy because of gravity or dark matter.

Albert Einstein in his theory of general relativity in 1915 found that gravity is not a force but a shape change in space/time caused by an uneven distribution of mass/energy. Large masses such as the earth warp space/time around it creating a "dip" into which the moon falls. Einstein must have watched birds in long distance flight. The lead bird in the "V" shaped flight pattern "warps" space/time allowing birds in back of the formation to draft on birds in front and thus conserving energy." Comparison can also be made to the NASCAR drivers or Bicycle racers who draft on the lead car or bicycle to increase speed and conserve energy (masses in motion affecting the motion of other masses).

DR. GEORGE WASHINGTON CARVER, "It is because of gravity that the giant sequoia tree grows perpendicular to the earth's core and not the earth's crust! The trunks of the Pine trees of the Rocky Mountains are perpendicular to the earth's core and not the mountainside where the tree is growing. No matter the topography of the land the tree trunk points to the earth's core. It is also because of gravity that the tiny lily droops. The lily essentially falls into the "warp" created by the mass of the earth (Einstein theory of general relativity)."

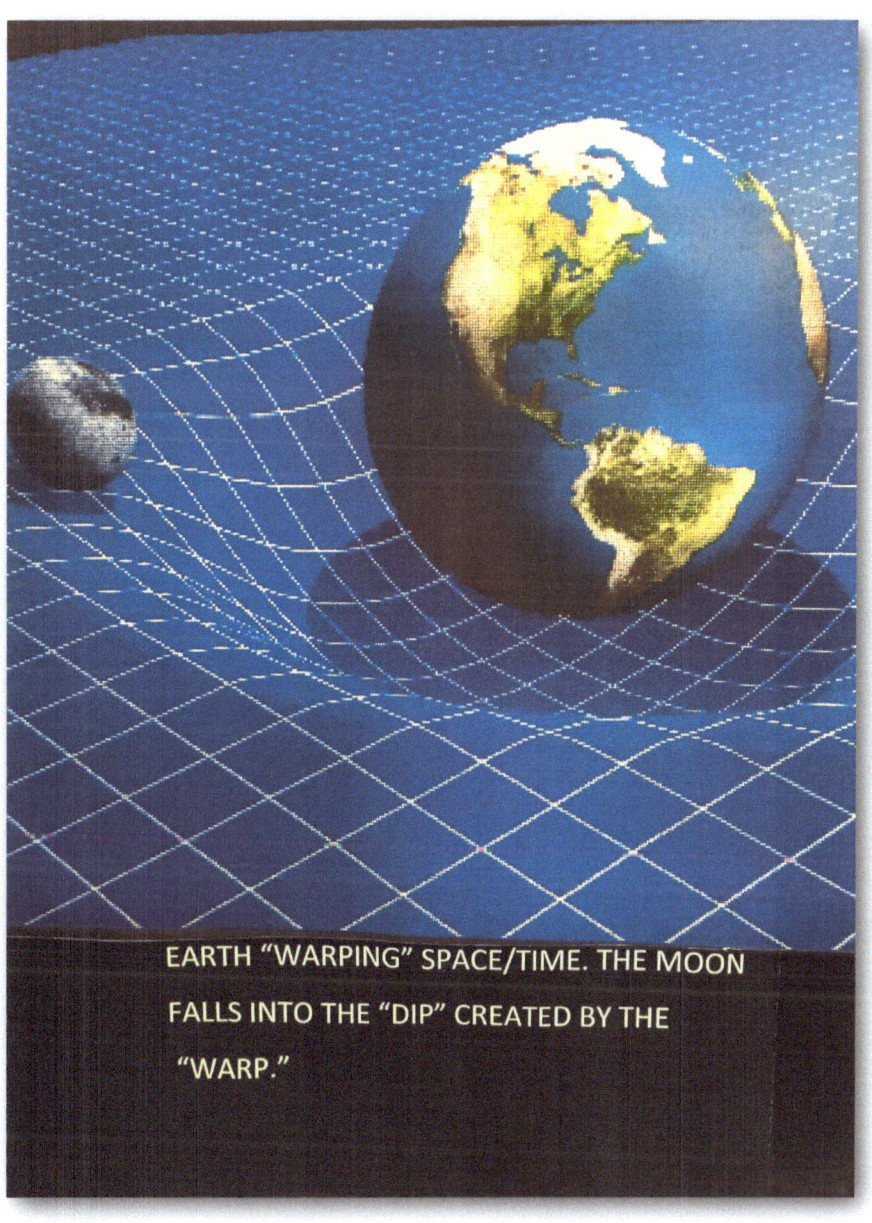

EARTH "WARPING" SPACE/TIME. THE MOON FALLS INTO THE "DIP" CREATED BY THE "WARP."

THE EARTH "WARPS" SPACE/TIME CAUSING THE LILLY TO FALL (DROOP) INTO THE "DIP" IN THE "WARPED" SPACE.

THE LEAD BIRD ALTERS SPACE/TIME SUCH
THAT BIRDS IN LINE HAVE AN EASIER PATH.

The center of gravity for planet Earth is at its core. Gravity weakens the further you get from the core. Trees grow perpendicular to the Earth's core rather than the irregular topography of the Earth's crust.

INDEX

• • •

ILLUSTRATIONS

· · ·

1. The earth contained in a grain of rice.
2. Botanist Dr. George Washington Carver stamp
3. Astronomer Galileo Galilei.
4. Artist concept of the **cosmic** "Big bang"
5. *Banksia Serrata* flower resembles **the** "Big Bang" with its red hot center and filaments radiating outward.
6. *Cephalanthus Occidentalis* flower (buttonbush) resembles the "Big Bang".
7. Sequoia tree seeds in the palm of the hand.
8. General Sherman Sequoia Tree in Sequoia National Park, California
9. Artist concept of the black hole with a bright hot round singularity at its base.
10. Artist concept of the black hole, event horizon and singularity.
11. Wine-cup (*Geissorhiza Radicans*) flower with the green singularity, indigo-blue accretion disk, ruby red black hole and white event horizon.
12. Wine-cup flower with all the elements of the black hole.
13. Hummingbird attracted to a *Trumpet vine* (*Campsis Radicans*) flower for its nectar reminds us of accreted matter entering the event horizon and being pulled into the black hole.

14. Universe expanding because of dark energy.

15. Water and photosynthesis are the energy of tree growth.

16. Tree trunk acting like dark matter holding the tree together while also facilitating the transport of water and products of photosynthesis energy inside the framework of the tree trunk.

17. Tree trunk holds the branches and leaves together just as Dark Matter holds the galaxies together.

18. Artist concept of dark matter providing the gravitational support for a cluster of galaxies.

19. Milky Way galaxy with rings.

20. Cut tree trunk with tree rings.

21. Orion constellation.

22. Orion nebula in the Orion constellation

23. Rosebud Nebula (NGC 7129) with more than 150 stars that are all younger than one million years old.

24. Celosia Argentea Spicata (cockscombs) flower resembles the Rosebud Nebula. Celosia in Greek meaning "burning."

25. Pillars of creation in the Eagle nebula where stars are born.

26. Opuntia Monacantha cactus (*Indian Fig Opuntia* or prickly pear cactus) where the substrate of cactus provides the foundation for the growth of fruits much as the cosmic nebula provides the substrate dust for the growth of stars.

27. Rose nebula (The Qur'an and the Oily Red Rose nebula)

28. Rose flower resembles the Rose Nebula.

29. Artist concept of the planet Mercury with its iron core.

30. Peach with large pit resembles the planet Mercury and its core.

31. Avocado with large pit resembles the planet Mercury and its core.

32. Artist conception of a comet with two tails *orbiting the sun.*

33. Computer –enhanced comet with two tails.

34. Actual comet with two tails

35. Maple tree with a seed that resembles the comet with two tails.

36. Earth warping space/time as described by Einstein in his theory of general relativity causing the moon to fall into the warp created by the earth and as a result, orbit.

37. Lily (*Lilium Candidum*) drooping to the earth in response to the warp in the earth's space /time gravity.

38. A "V"- formation of migrating birds (sometimes called a SKEIN) improves efficiency of flight. All birds except the first fly in the upwash from one of the wingtip vortices of the bird ahead. The upwash assists each bird in supporting its own weight. This is an example of how one mass in motion can affect another mass in motion.

39. Mountain pine trees growing perpendicular to the earth's core. The earth's gravity is centered at its core. Trees respect the law of gravity. Notice that the tree does not grow perpendicular to the earth's crust.

www.ingramcontent.com/pod-product-compliance
Lightning Source LLC
Chambersburg PA
CBHW050742180526
45159CB00003B/1315